Computational
Fluid Dynamics
Laboratory Manual

Computational Fluid Dynamics
Laboratory Manual

M Veeramanikandan ME, PhD
Assistant Professor
Department of Mechanical Engineering
Sri Ramakrishna Institute of Technology
Anna University
Pachapalayam, Perur Chettipalayam
Coimbatore, Tamil Nadu

D Sathish ME, PhD
Assistant Professor
Department of Mechanical Engineering
Sri Ramakrishna Institute of Technology
Anna University
Pachapalayam, Perur Chettipalayam
Coimbatore, Tamil Nadu

CBS

CBS Publishers & Distributors Pvt Ltd
New Delhi • Bengaluru • Chennai • Kochi • Kolkata • Mumbai
Hyderabad • Jharkhand • Nagpur • Patna • Pune • Uttarakhand

Computational Fluid Dynamics
Laboratory Manual

ISBN: 978-93-90046-06-5

Copyright © Authors and Publisher

First Edition: 2021

Published by Satish Kumar Jain and produced by Varun Jain for
CBS Publishers & Distributors Pvt Ltd
4819/XI Prahlad Street, 24 Ansari Road, Daryaganj, New Delhi 110 002, India.
Ph: 011-23289259, 23266861, 23266867   Fax: 011-23243014
Website: www.cbspd.com   e-mail: delhi@cbspd.com; cbspubs@airtelmail.in.

Corporate Office: 204 FIE, Industrial Area, Patparganj, Delhi 110 092, India
Ph: 011-4934 4934   Fax: 011-4934 4935   e-mail: publishing@cbspd.com; publicity@cbspd.com

Branches

- **Bengaluru:** Seema House 2975, 17th Cross, K.R. Road, Banasankari 2nd Stage, Bengaluru 560 070, Karnataka, India
  Ph: +91-80-26771678/79   Fax: +91-80-26771680   e-mail: bangalore@cbspd.com
- **Chennai:** 7, Subbaraya Street, Shenoy Nagar, Chennai 600 030, Tamil Nadu, India
  Ph: +91-44-26680620, 26681266   Fax: +91-44-42032115   e-mail: chennai@cbspd.com
- **Kochi:** 42/1325, 1326, Power House Road, Opposite KSEB, Power House, Ernakulum-682018, Kochi, Kerala, India
  Ph: +91-484-4059061-67   Fax: +91-484-4059065   e-mail: kochi@cbspd.com
- **Kolkata:** 6/B, Ground Floor, Rameswar Shaw Road, Kolkata-700 014 (West Bengal), India
  Ph: +91-33-22891126, 22891127, 22891128   e-mail: kolkata@cbspd.com
- **Mumbai:** 83-C, Dr E Moses Road, Worli, Mumbai-400018, Maharashtra, India
  Ph: +91-22-24902340/41   Fax: +91-22-24902342   e-mail: mumbai@cbspd.com

Representatives

Hyderabad	0-9885175004	**Jharkhand**	0-9811541605	**Nagpur**	0-9421945513
Patna	0-9334159340	**Pune**	0-9623451994	**Uttarakhand**	0-9716462459

Printed at Rashtriya Printers, Dilshad Garden, Delhi, India

Preface

Computational fluid dynamics is an important tool to investigate fluid flow problems in industry and academia. This course can be taken without a prior background in computational techniques, however, a background of fundamental fluid dynamics, partial differential equations, linear algebra and programming language is desirable. The primary focus of this course is to gain a solid foundation of numerical methods for different fluid problems like convection–diffusion problems. The emphasis is on the physical meaning underlying the focused mathematics. Control volume method, which is a robust physically intuitive numerical approach, widely used in industry and academia alike, described alongwith its applications in various fields.

We express our heartfelt gratitude to the management of Sri Ramakrishna Institute of Technology, Coimbatore, for providing support in carrying out this work.

M Veeramanikandan
D Sathish

Contents

Experiment 1

One-Dimensional Steady State Diffusion

AIM

To solve the one-dimensional steady state diffusion for a given problem using Workbench and Fluent to study the results.

PROBLEM SPECIFICATION

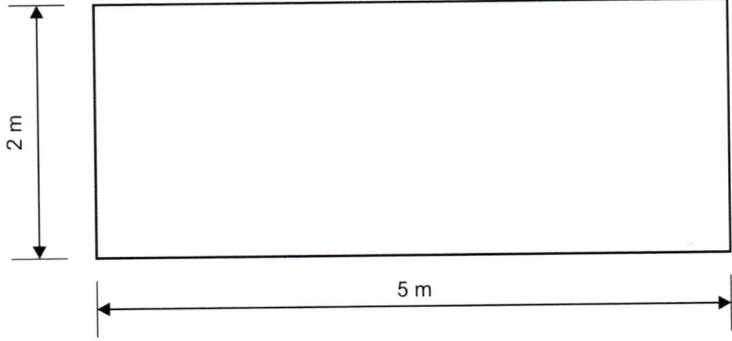

Consider a rectangular section having width W = 5 m and height H = 2 m. The following conditions are applied to analyse this problem.

	Case 1	Case 2	Case 3
Left wall	Temperature = 600 K	Heat flux = 500 W/m²	Heat flux = 500 W/m²
Right wall	Temperature = 300 K	Temperature = 300 K	Convective heat transfer coefficient = 25 W/m²K Temperature = 300 K

PROCEDURE

Preprocessor

All the steps are carried out using Workbench.

Model Creation

1. Vertices are created using (0,0), (5,0), (5,2) and (0,5)
2. Edges are created using the above vertices
3. Select all the edges and convert them to faces

Meshing

1. Horizontal top and bottom edges are meshed
2. Vertical left and right side edges are meshed

Specification of Boundary Types

Edge	Name	Type
Left	Left wall	Wall
Right	Right wall	Wall
Top	Top wall	Wall
Bottom	Bottom wall	Wall

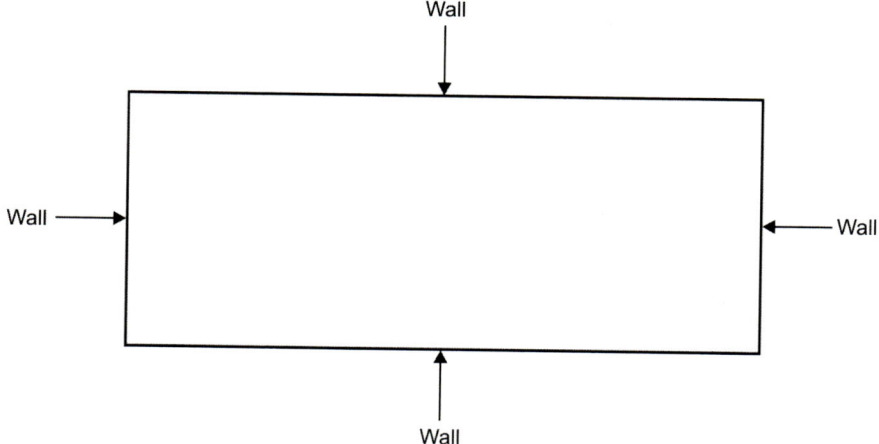

1. The continuum is specified as solid
2. Export the mesh file with export 2D mesh option select
3. The mesh file has been saved as plate mesh

Processor

1. The meshed file is read to Fluent case file
2. Energy equation is selected as this problem involves in temperature distribution (Laplace equation is solved $K\,(d^2T/dx^2) = 0$)
3. Steel has been selected as material

Boundary Conditions

1. Temperature of 600 K is applied at the left edge and 300 K is applied at the right edge (case I)

2. The solution is initialised from inlet
3. The solution is obtained by iterative process and is converged

Postprocessor

1. The graph is plotted between length of the plate on x-axis and the temperature at the wall on y-axis
2. Then the boundary conditions are changed in processor and postprocessor are repeated
3. Comparative graphs are drawn for various boundary conditions

Inference

From the graph, it is evident that the temperature decreases consistently from high temperature side.

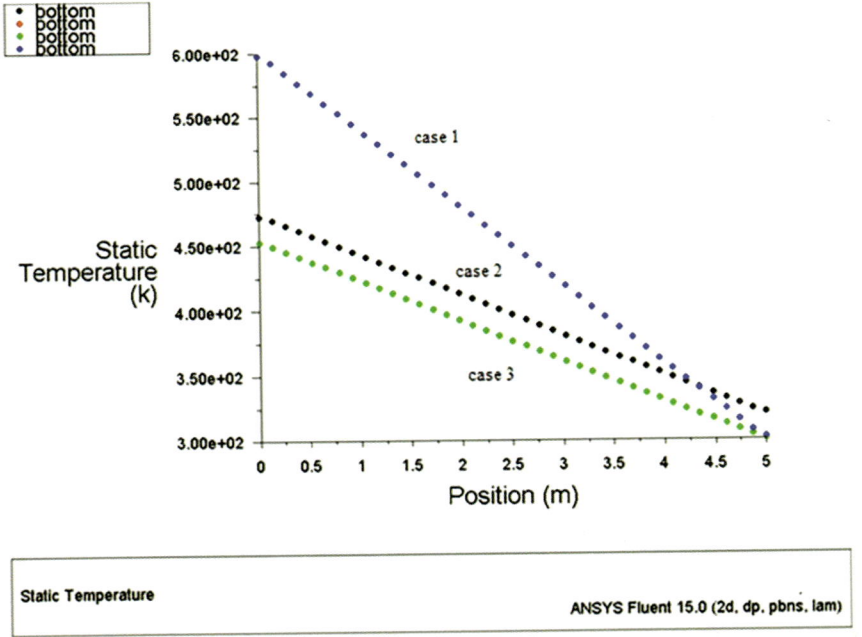

RESULT

The given problem is solved using Workbench and Fluent and the results are drawn for different boundary conditions.

Experiment 2

One-Dimensional Steady State Diffusion with Volume Source

AIM

To solve one-dimensional steady state diffusion with volume source for a given problem using Workbench and Fluent to study the results.

Problem Specification

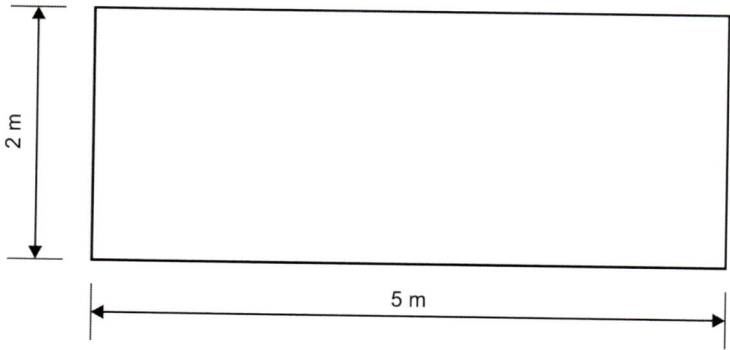

Consider a rectangular section having width (W) = 5 m and height (H) = 2 m. The following conditions are applied to analyse the problem.

	Case 1	Case 2	Case 3
Left wall	Temperature = 600 K	Heat flux = 500 W/m²	Convective heat transfer coefficient = 25 W/m²K Temperature = 300 K
Right wall	Temperature = 300 K	Convective heat transfer coefficient = 25 W/m²K Temperature = 300 K	Convective heat transfer coefficient = 25 W/m²K Temperature = 300 K
Source heat = 1000 W/m³.			

PROCEDURE

Preprocessor

All the steps are carried out using Workbench

Model Creation

1. Vertices are created using (0,0), (5,0), (5,2) and (0,2)
2. The edges are created using the vertices created
3. The edges are converted to faces by selecting all the edges

Meshing

1. Horizontal top and bottom edges are meshed
2. Vertical left and right side edges are meshed

Specification of Boundary Types

Edge	Name	Type
Left	Left wall	Wall
Right	Right wall	Wall
Top	Top wall	Wall
Bottom	Bottom wall	Wall

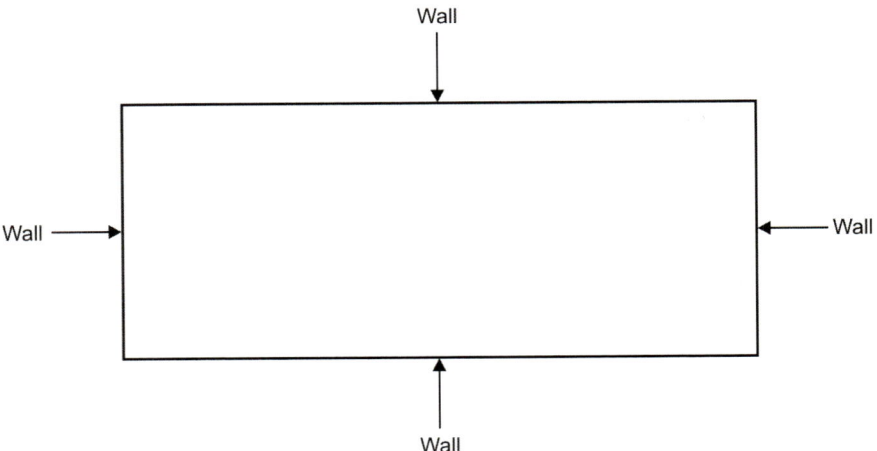

1. The continuum is specified as solid
2. Export the mesh file with export 2D mesh option select
3. The mesh file has been saved as plate mesh

Processor

1. The meshed file is read to Fluent case file
2. Energy equation is selected as this problem involves in temperature distribution
   (Poisson's equation is solved, i.e. $K(d^2T/dx^2 + Q) = 0$)
3. Steel has been selected as material

Boundary Conditions

1. Temperature of 600 K is applied at the left edge and 300 K is applied at the right edge
2. The solution is initialised from the left wall
3. The solution is obtained by iterative process and is converged

Postprocessor

1. The graph is plotted between length of the plate on x-axis and the temperature at the wall on y-axis
2. Then the boundary conditions are changed in processor and postprocessor is repeated
3. Comparative graphs are drawn for various boundary conditions

Inference

From the graph, the temperature decreases consistently from high temperature side.

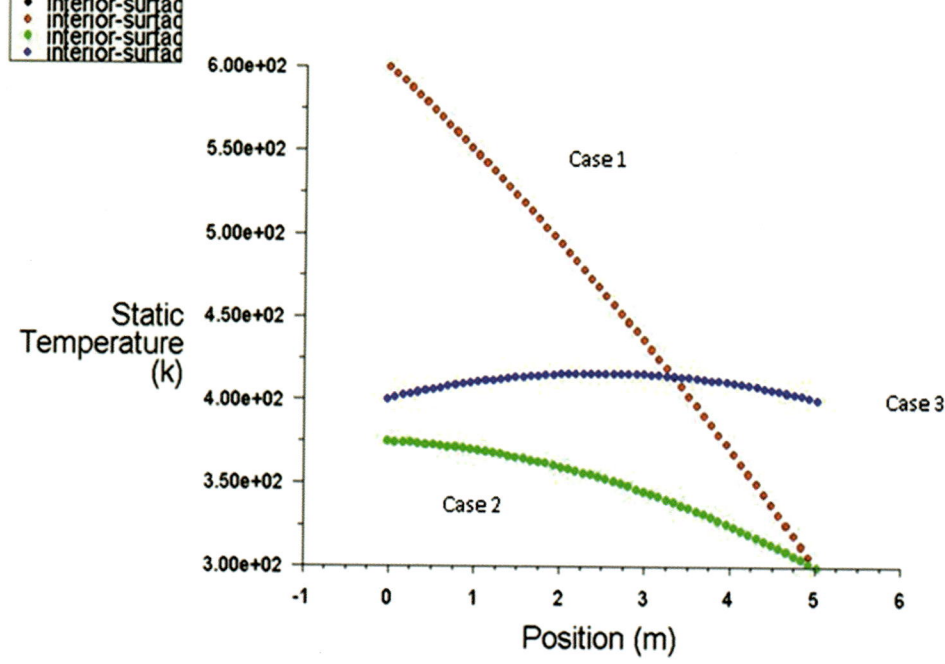

RESULT

The given problem is solved using Workbench and Fluent and the results are drawn for different boundary conditions.

Experiment 3

Heat Transfer in a Circular Fin

AIM

To solve heat transfer in a circular fin using Workbench and Fluent to study the result.

Problem Specification

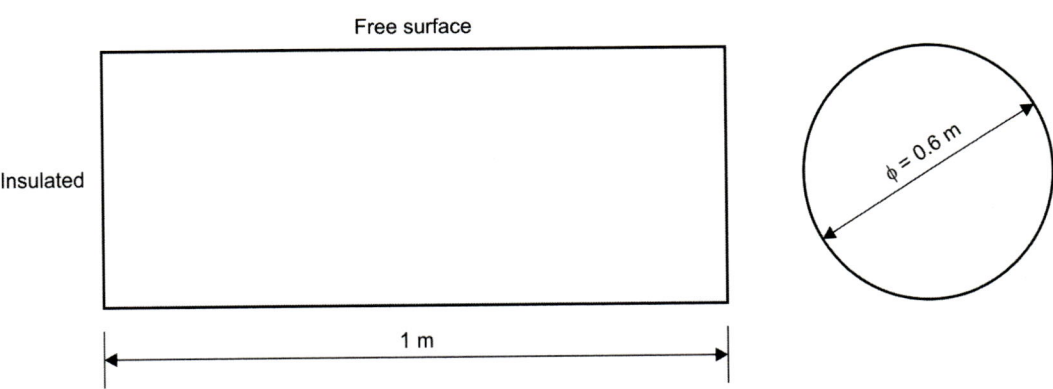

A circular fin of 0.2 m diameter and 1 m length is considered, the left wall is subjected to temperature, the right wall is insulated and the top wall is subjected to atmospheric condition.

	Case 1	Case 2	Case 3
Left wall	Temperature = 473 K	Temperature = 500 K	Temperature = 700 K
Top wall	Free stream temperature = 298 K Convection coefficient $h = 15$ W/m²	Free stream temperature = 303 K Convection coefficient $h = 15$ W/m²	Free stream temperature = 298 K Convection coefficient $h = 20$ W/m²
Thermal conductivity	$K = 6$ W/mK	$K = 10$ W/mK	$K = 12$ W/mK

PROCEDURE

Preprocedure

All the steps are carried out using Workbench

Model Creation

1. Vertices are created using (0,0), (1,0), (1,0.1) and (0,0.1)
2. Edges are created using vertices
3. Edges are converted to faces by selecting all the edges

Meshing

1. Horizontal and vertical edges are meshed into 50 divisions and 1 division respectively
2. Face mesh has been done using face mesh by selecting the face

Specification of Boundary Types

Edge	Name	Type
Left	Left wall	Wall
Right	Right wall	Wall
Top	Top wall	Wall
Bottom	Center line	Axis

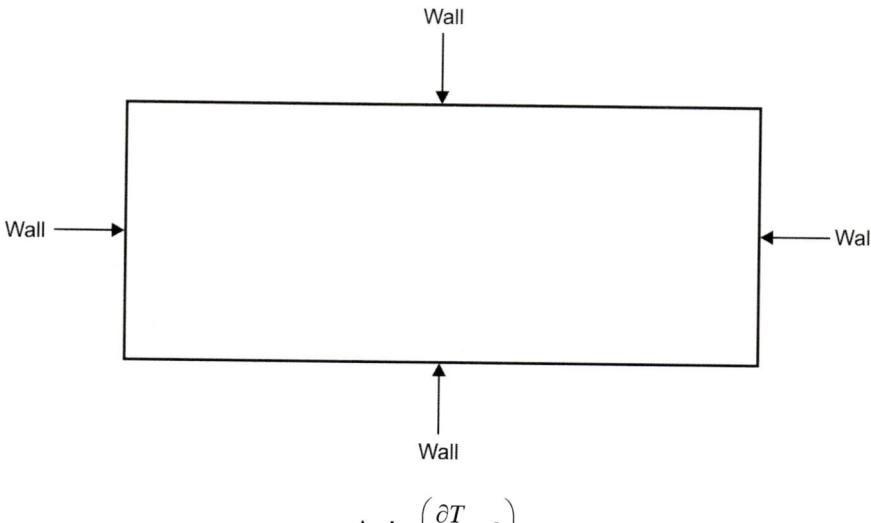

1. Continuum is specified as solid
2. Export the mesh file with 2D mesh option
3. Mesh file has been saved as fin mesh

Procedure

1. The meshed file is read to Fluent as case file
2. Energy equation is selected as this problem involves in temperature distribution and axisymmetric
3. Aluminium has been selected as the material K = 6 W/mK

Boundary Conditions

1. Temperature of 473 K is applied to the left wall
2. Free stream temperature of 298 K and convective heat transfer coefficient 15 W/wm² at top wall is applied
3. Temperature of right wall is zero
4. The solution is initialized from the left wall
5. Then the solution is obtained by iterative process and is converged

Postprocedure

1. The graph is plotted between length of the fin on x-axis and the temperature at the top wall on y-axis
2. The boundary conditions are changed in processor and postprocessor are repeated
3. Comparative graphs are drawn for various boundary conditions

THEORETICAL RESULT

Perimeter
$$P = 2\pi r$$
$$= 2 \times \pi \times 0.1$$
$$= 0.6283 \text{ m}$$

Area
$$A = \pi r^2$$
$$= \pi \times 0.1^2$$
$$= 0.0314 \text{ m}^2$$

Case 1:

$$h = 15 \text{ W/m}^2 \qquad [K = 6 \text{ W/mK}]$$

$$n^2 = \frac{15 \times 0.6283}{6 \times 0.0314}$$

$$= 50.024$$

$$n = 7.073$$

$$(T_x - T_{atm})/(T_L - T_{atm}) = \frac{\cos h\left[n(L-x)\right]}{\cos h(nL)}$$

$$\frac{T_x - 298}{473 - 298} = \frac{\cos h\left[7.073(1-0.5)\right]}{\cos h(7.073 \times 1)}$$

$$T_x = 303.1 \text{ K}$$

Inference

- From the graph at x= 0.5 m
- Temperature   = 305 K
- Theoretical value = 303.1 K

Both the results are almost equal

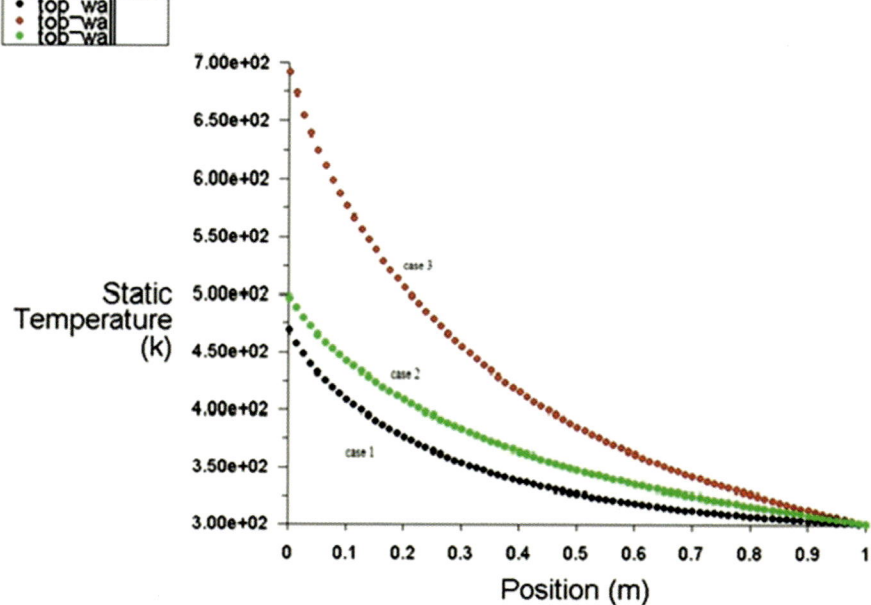

RESULT

The problem is solved using Workbench and Fluent and the results are compared with theoretical result.

Experiment 4

One-Dimensional Transient Heat Conduction

AIM

To solve one-dimensional transient heat conduction for a given problem using Workbench and Fluent to study the result.

Problem Specification

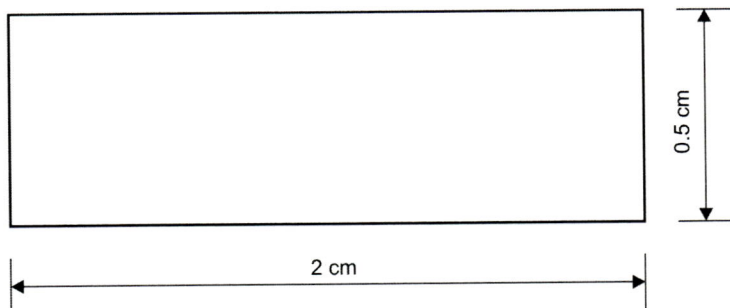

- A plate of 2 cm × 0.5 cm size has initial temperature of 473 K and left side the temperature is 273 K, all the other walls are insulated
- The plate material is aluminium with density ρ = 5000 kg/m³ and specific heat C_p = 1000 J/kg K and thermal conductivity, K = 10 W/mK

PROCEDURE

Preprocessor

All the steps are carried out using Workbench

Model Creation

1. Vertices are created using (0,0), (2,0), (2,0.5) and (0,0.5) coordinates
2. Edges are created using the vertices
3. Faces are created by selecting all the edges

Meshing

1. Horizontal and vertical edges are meshed into 10 divisions and 1 division respectively
2. Face mesh is performed

Boundary Type Specification

Edge	Name	Type
Left	Left wall	Wall
Right	Right wall	Wall
Top	Top wall	Wall
Bottom	Bottom wall	Wall

1. The continuum is selected as solid
2. The mesh file is exported with export 2D mesh option selected
3. The mesh file is saved as plate mesh

Processor

1. The mesh file is read as a case file
2. The problem is selected as unsteady in the solver and the energy equation to find the temperature distribution for a required time is

$$\rho C_p \frac{\partial T}{\partial t} = K \frac{\partial^2 T}{\partial t^2}$$

Boundary Conditions

1. The left wall are subjected to 273 K temperature, top and bottom to 0 K temperature
2. Material properties are changed as given in the problem
3. The solution is initialized from all zones where the initial temperature is applied as 473 K
4. The solution is converged by iterating process

Postprocessor

1. The graph is plotted between the length of plate on the x-axis and the static temperature at the top wall on y-axis
2. The solution is iterated for different time intervals
3. Comparative results are drawn

Inference

1. Right side temperature decreases with increase in time
2. At 275 seconds, the condition is changed to a steady state

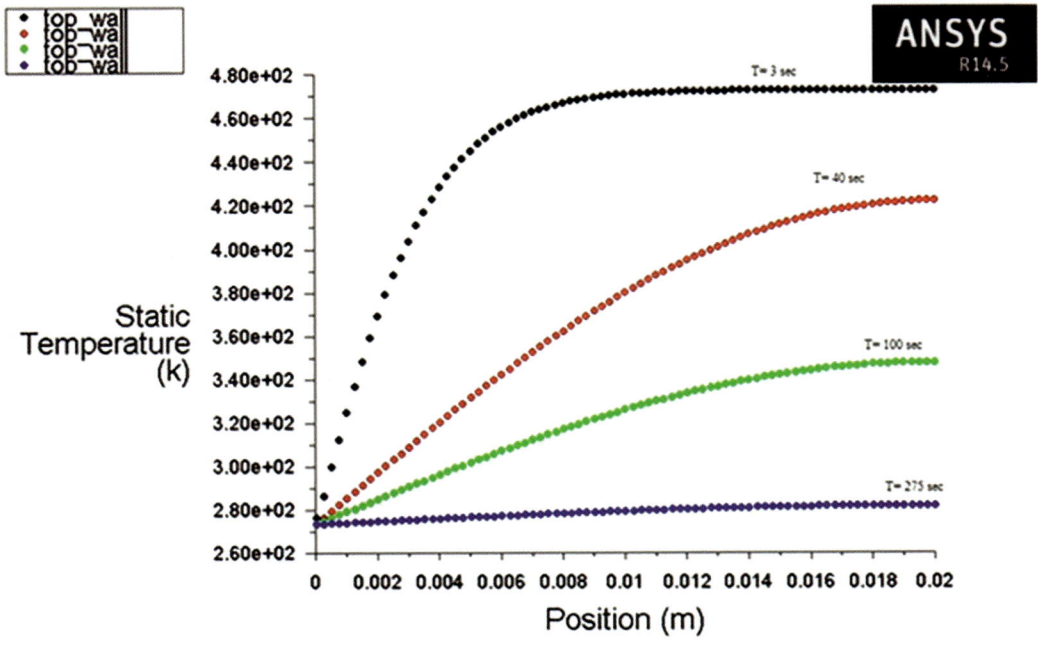

RESULT

The given problem is solved using Workbench and Fluent and result are drawn for various time intervals.

Experiment 5

Laminar Pipe Flow

AIM

To solve the given laminar pipe flow problem using ANSYS Fluent and Fluent to study the results

Problem Specification

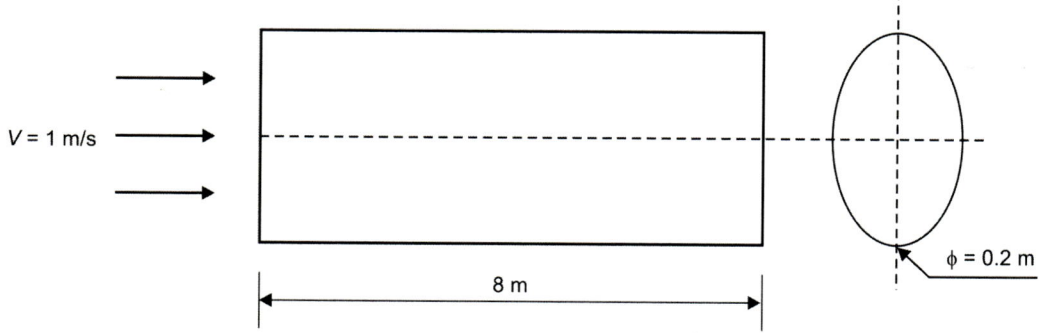

- Pipe diameter of 0.2 m × 8 m length is considered through which the fluid is flowing with a velocity of 1 m/s
- The fluid exhausts into the ambient condition at 1 atm. Take density ρ = 1 kg/m³ and coefficient of viscosity η = 2 × 10⁻⁵ kg/m³

PROCEDURE

Preprocessor

All the steps in the preprocessor are using ANSYS Fluent

Model Creation

1. Vertices are created using coordinates (0,0), (8,0), (8,0.1) and (0,0.1)
2. Edges are created using created vertices
3. The faces are created using the created edges

Meshing

1. Horizontal top and bottom edges are selected and meshed into 100 divisions
2. Vertical left and right edges are selected and meshed into 5 divisions
3. Face mesh is performed

Boundary Type Specification

Edge	Name	Type
Left	Inlet	Velocity-inlet
Right	Outlet	Pressure-outlet
Top	Wall	Wall
Bottom	Center line	Axis

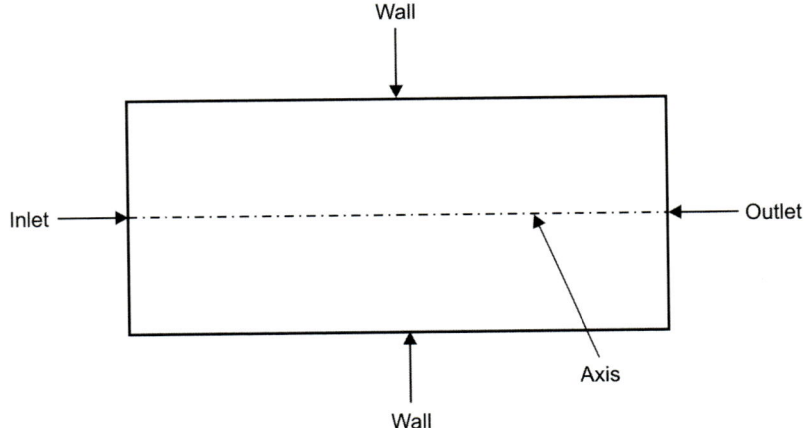

1. The continuous is selected as fluid
2. The mesh is exported with export 2D mesh option selected
3. The mesh file has been saved as pipe mesh

Processor

1. The meshed file is read to Fluent as case file
2. The problem has been taken as axis-symmetric
3. Material properties are changed to density $\rho = 1 \text{ kg/m}^3$
4. Viscosity $\eta = 2 \times 10^{-3} \text{ kg/ms}$
5. Operating pressure of 10325 Pascal is fixed

Boundary Conditions

1. Inlet velocity of 1 m/s is applied at the left edge
2. Outlet pressure at the right edge is applied as 0
3. Top edge of the model has been taken as adiabatic wall
4. Momentum equation is changed to second order upwind scheme from first upwind

5. The solution is initialized from inlet
6. Then the solution is obtained by iterative process and converged

Postprocessor

1. The graph 'A' is plotted between length of the pipe on x-axis and axial velocity at the axis on y-axis
2. The graph 'B' is plotted between length of the pipe on x-axis and skin friction coefficient at the wall on y-axis
3. The graph 'C' is plotted between radius of the pipe on x-axis and the axial velocity at the outlet on y-axis
4. Then the mesh is refined to 100 × 10 and 100 × 20 in preprocessor and all the steps are repeated
5. Comparative graphs are drawn for various grid sizes

Inference

1. The velocity at the center line in the fully developed region for the thinner mesh is 1.98 m/s
2. Skin friction coefficient of 0.159 is obtained in the fully development region

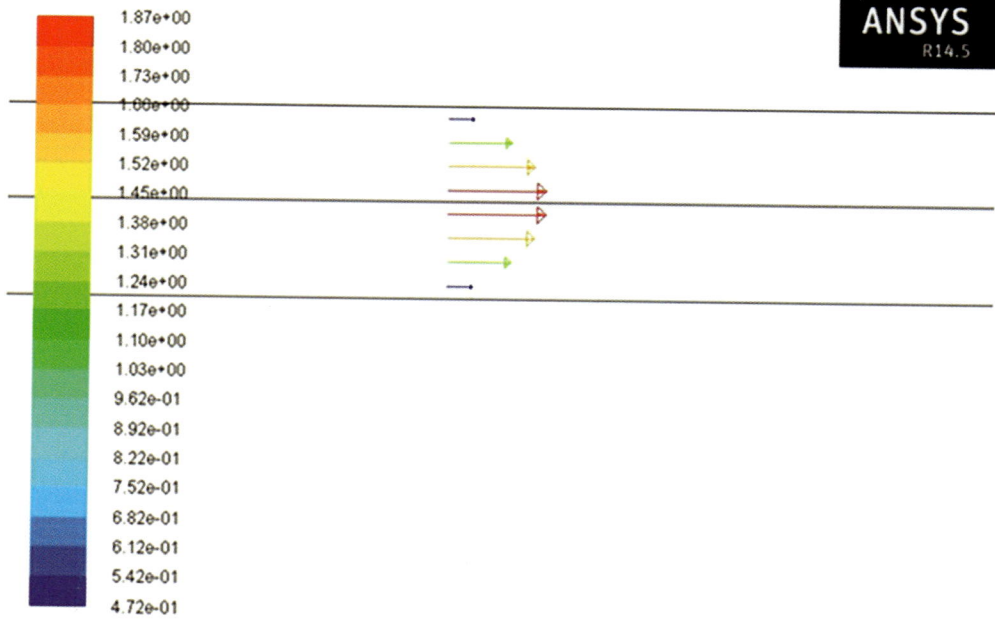

Velocity Vectors Colored By Velocity Magnitude (m/s)

ANSYS Fluent 14.5 (axi, dp, pbns, lam)

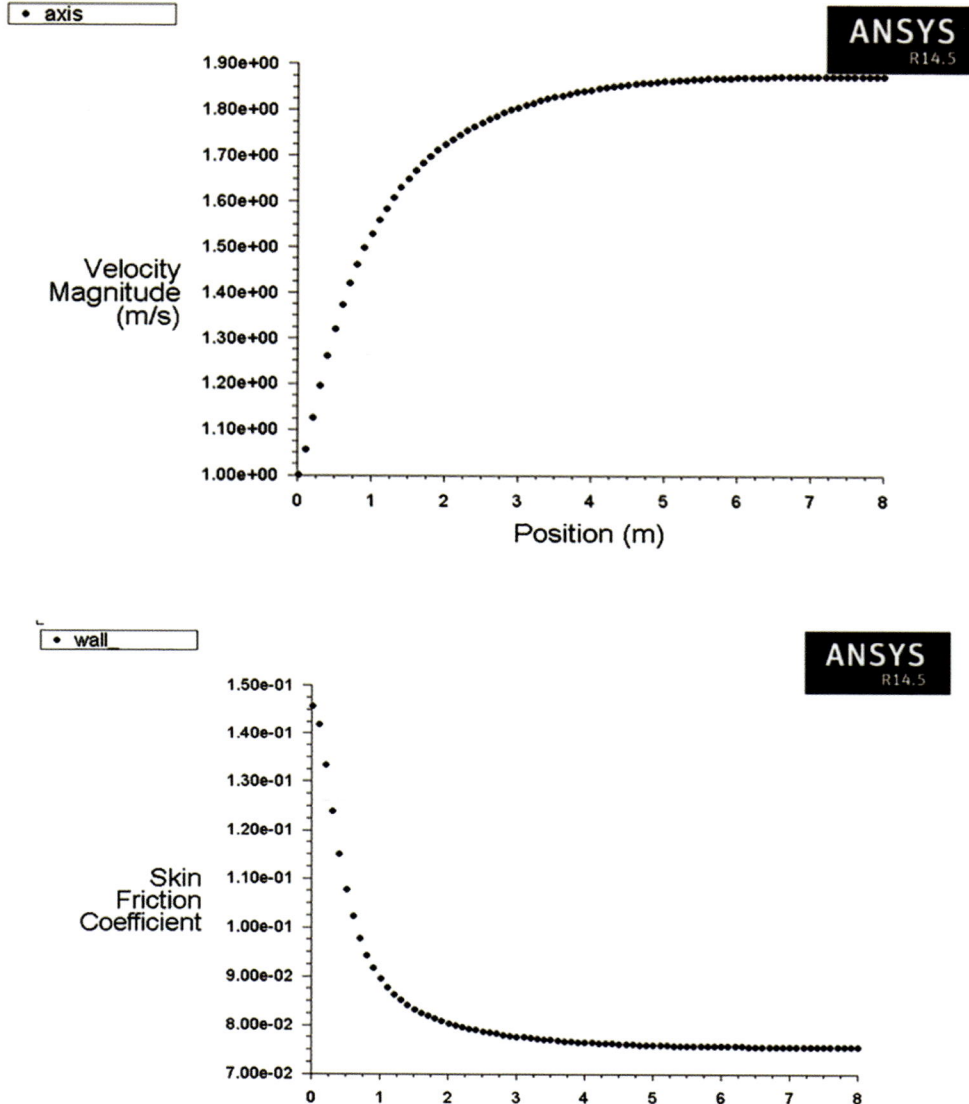

RESULT

The laminar flow problem is solved by ANSYS Fluent, the velocity and skin friction coefficient graphs are plotted.

Experiment 6

Turbulent Pipe Flow

AIM

To solve the given turbulent pipe flow problem using ANSYS Fluent and Fluent the given to study results

Problem Specification

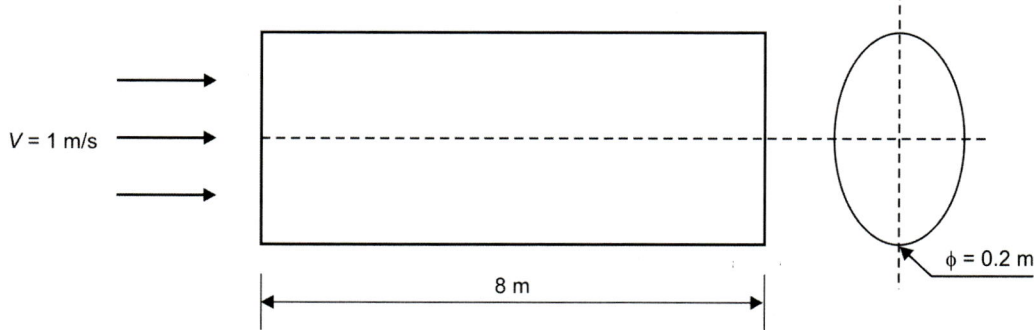

- The pipe diameter of 0.2 m × 8 m length is considered through which the fluid is flowing with a velocity of 1 m/s
- The fluid exhausts into ambient condition which is at 1 atm. Take density $\rho = 1$ kg/m³ and coefficient of viscosity $\eta = 2 \times 10^{-5}$ kg/m³

PROCEDURE

Preprocessor

All the steps in the preprocessor are carried out using ANSYS Fluent.

Model Creation

1. Vertices are created using coordinates (0,0), (8,0), (8,0.1) and (0,0.1)
2. Edges are created using created vertices
3. The faces are created using the created edges

Meshing

1. Horizontal top and bottom edges are selected and meshed into 100 divisions
2. Vertical left and right edges are selected and given first line as 0.001 and meshed into 30 divisions. The meshes are closer towards the wall. It is performed to predict boundary layer separation (Y plus value)
3. Face mesh is performed

Specification of Boundary Types

Edge	Name	Type
Left	Inlet	Velocity-inlet
Right	Outlet	Pressure-outlet
Top	Wall	Wall
Bottom	Center line	Axis

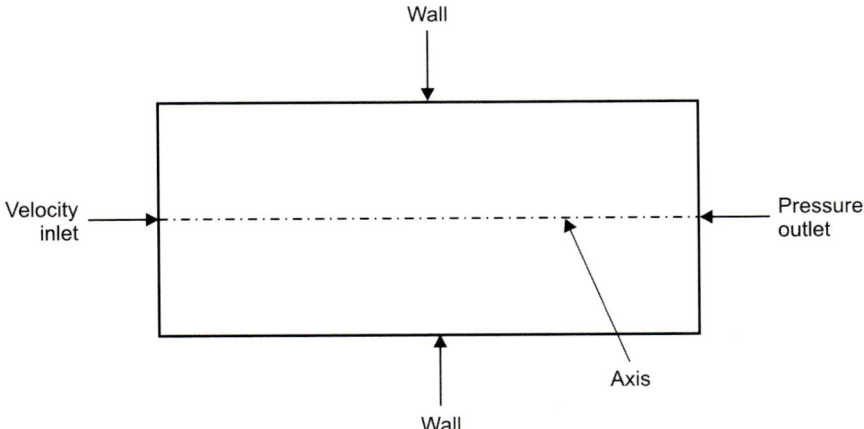

1. The continuum is selected as fluid
2. The mesh is exported with export 2D mesh option selected
3. The mesh file has been saved as pipe mesh

Processor

1. The meshed file is read to Fluent as case file
2. The problem has been taken as axis-symmetric and turbulent, *k*-epsilon viscous flow model, the enhanced wall treatment option is selected
3. Density $\rho = 1$ kg/m^3 and viscosity $\eta = 2 \times 10^{-5}$ kg/ms are applied to the material
4. Operating pressure of 101325 Pascal is fixed

Boundary Conditions

1. Inlet velocity of 1 m/s is applied at the edge and select intensity and hydraulic diameter under
   a. Turbulence intensity = 1%
   b. Hydraulic diameter = 0.2 m
2. Outlet pressure at the right edge is applied as 0

3. Top edge of the model has been taken as adiabatic wall
4. Momentum equation, turbulence kinetic energy dissipation rate equations are changed to second upwind scheme from first order upwind scheme
5. The solution is initialized from inlet
6. Then the solution is obtained by iterative process and is converged

Postprocessor

1. The graph 'A' is plotted between length of the pipe on x-axis and y plus value at the wall on y-axis
2. The graph 'B' is plotted between length of the pipe on x-axis and axial velocity at the axis on y-axis
3. The graph 'C' is plotted between length of the pipe on x-axis and the skin friction coefficient at the wall on y-axis
4. The graph 'D' is plotted between radius of pipe on x-axis and the axial velocity at the outlet on y-axis
5. The mesh of 100 × 60 is refined in ANSYS Fluent and all the steps are repeated
6. Comparative graphs are drawn for various grid sizes

Inference

Skin friction coefficient remains constant at the fully developed region.

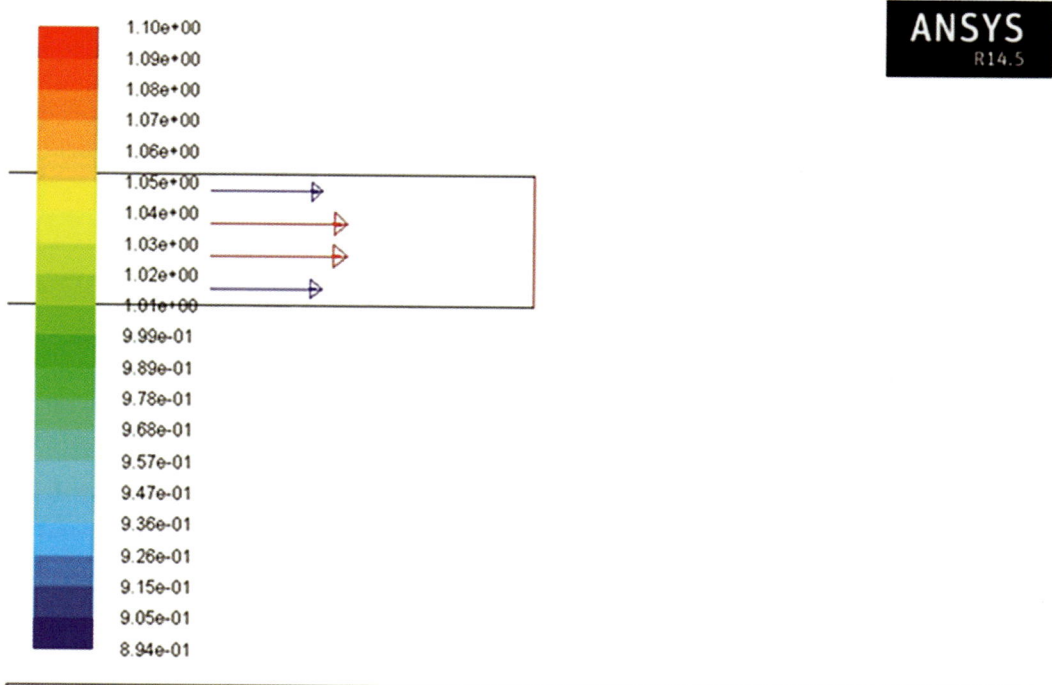

Velocity Vectors Colored By Velocity Magnitude (m/s)

ANSYS Fluent 14.5 (2d, dp, pons, ske)

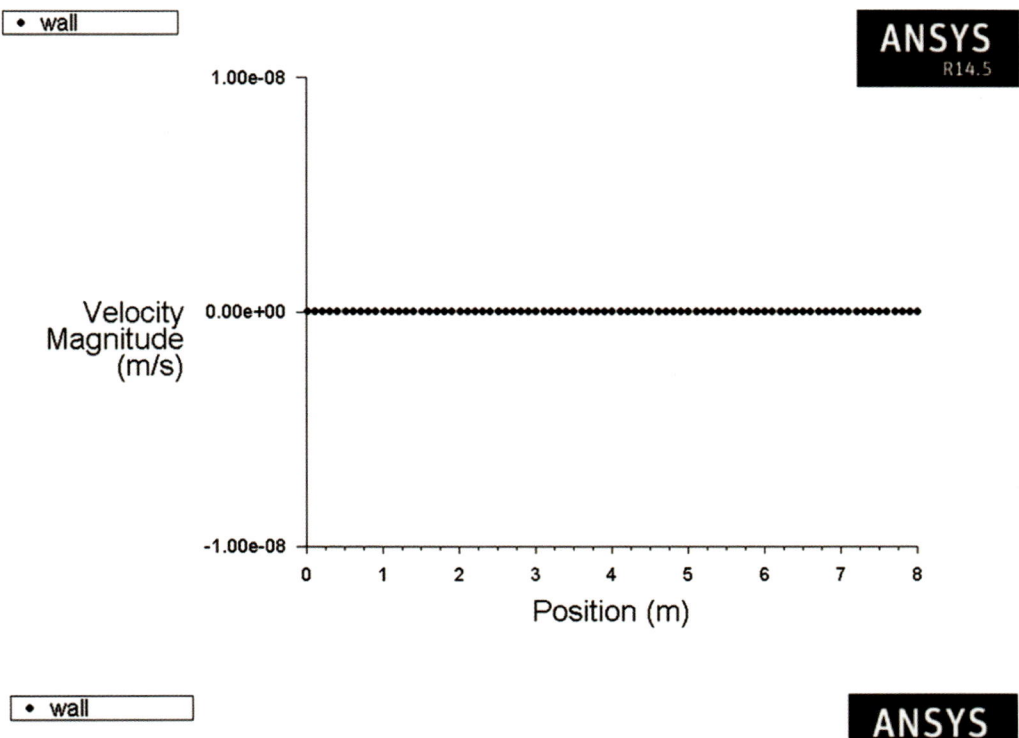

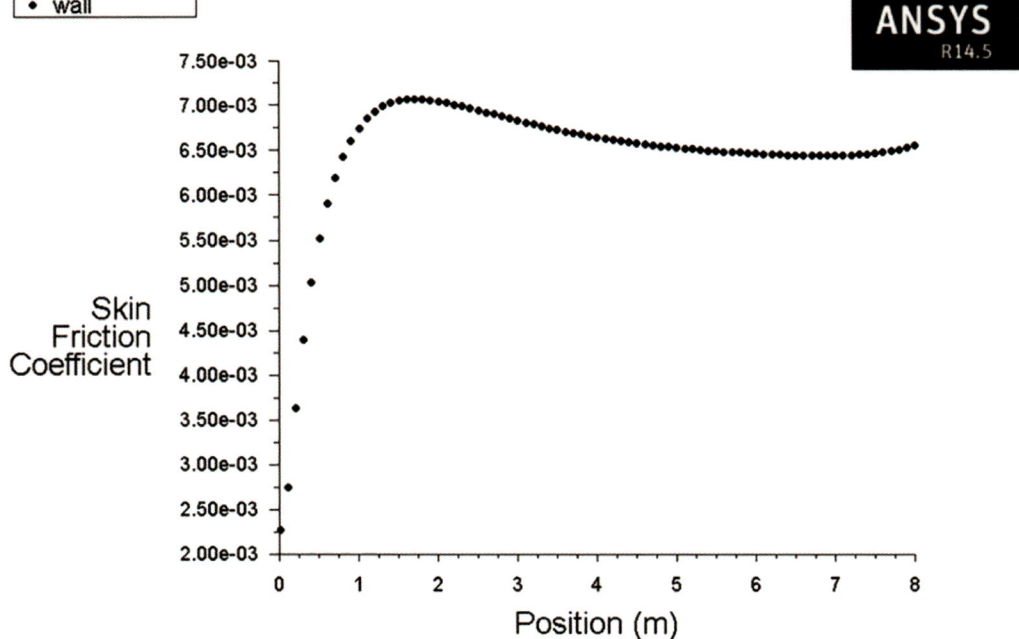

RESULT

The given problem is solved using ANSYS Fluent and Fluent and the result is drawn for different grid sizes.

External Flow
over a Flat Plate

AIM

To solve the external flow over a flat plate problem by using Workbench and Fluent to study the result.

Problem Specification

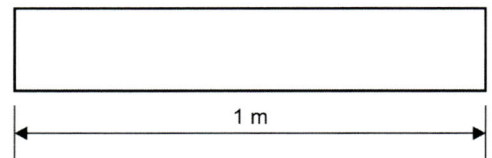

1 m

A plate of 1 m × 0.5 m size is taken the left wall is subjected to velocity and right wall to pressure.

PREPROCESSOR

Model Creation

1. Vertices are created using (0,0), (1,0), (1,0.5) and (0,0.5) coordinates
2. Edges are created using the vertices
3. Faces are created by selecting all the edges

Meshing

1. Horizontal and vertical edges are meshed into 50 divisions and 1 division respectively
2. Face mesh is performed

Specification of Boundary Types

Edge	Name	Type
Left	Left wall	Wall
Right	Right wall	Wall
Top	Top wall	Wall
Bottom	Bottom wall	Wall

1. The continuum is selected as solid
2. The mesh file is exported with export 2D mesh option selected
3. The mesh file is saved as plate mesh

Procedure

1. The meshed file is read to Fluent as case file
2. Laminar is selected, for turbulent flow *k*-epsilon is selected
3. Aluminium has been selected as the material

Boundary Conditions

1. Velocity of 5 m/s is applied to the left wall
2. The solution is initialized from the left wall
3. Then the solution is obtained by iterative process and is converged

Postprocessor

1. The graph is plotted
2. The solution is iterated for different time intervals
3. Comparative results are drawn

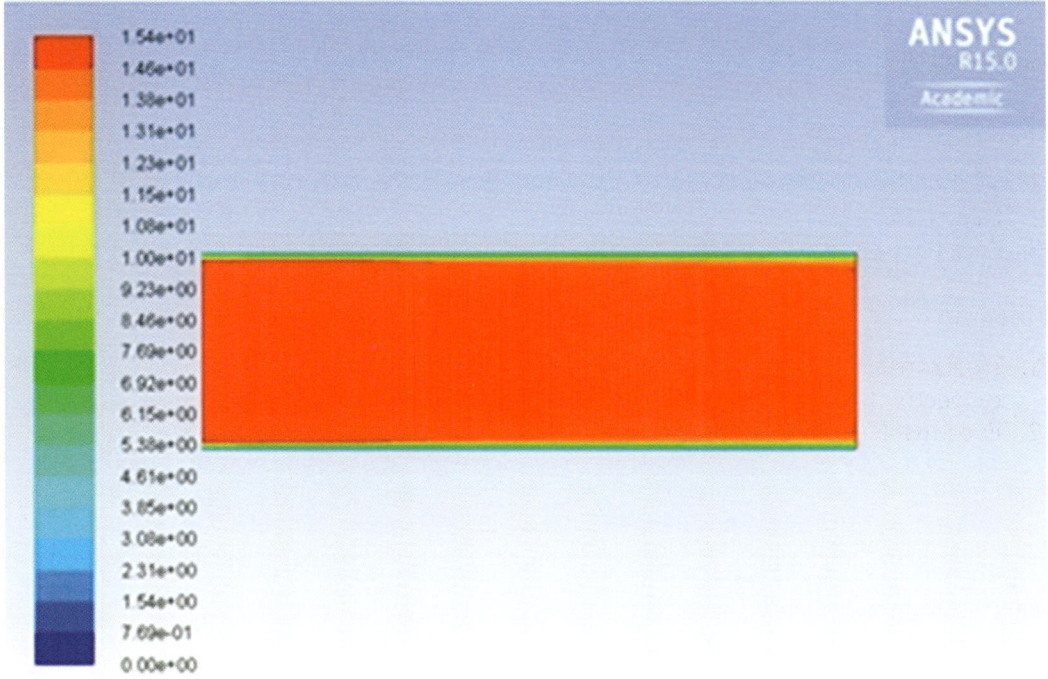

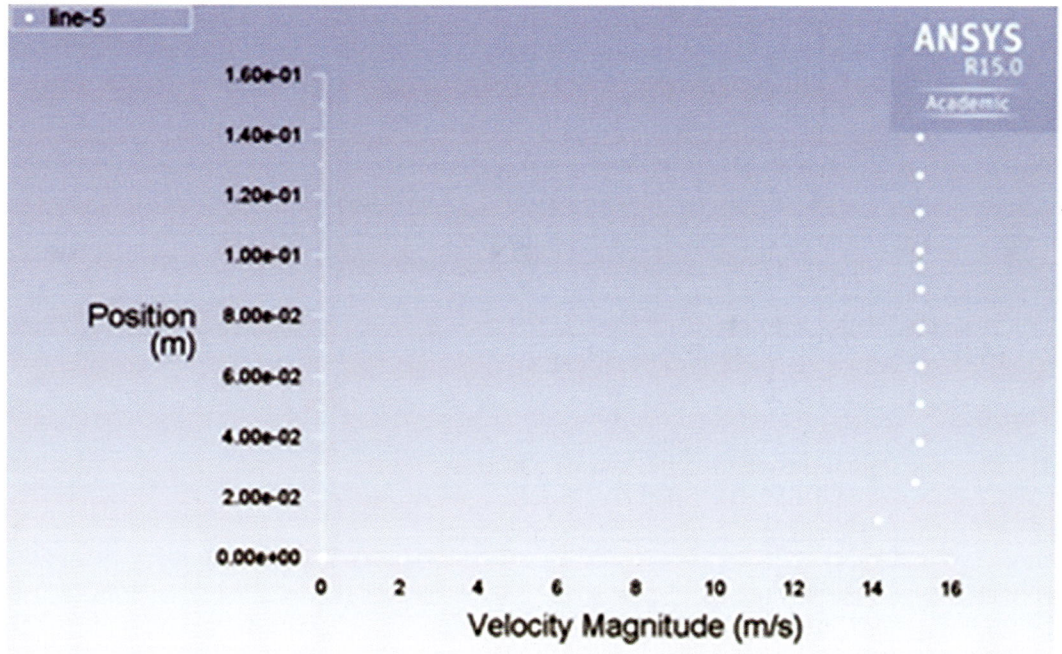

RESULT

The given problem is solved using Workbench and Fluent.

Experiment 8

Flow over a Cylinder

AIM

To solve flow over a cylinder for a given problem using Workbench and Fluent to study the result.

Problem Specification

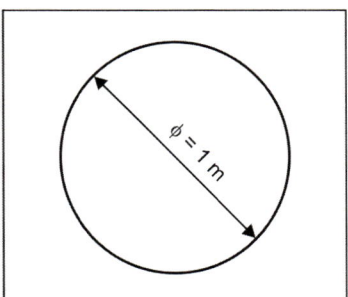

PROCEDURE

Preprocessor

Model Creation

1. Vertices are created using (−3,2), (−3, −2), (5,2) and (5,-2) coordinates
2. Edges are using the created vertices
3. Face is created by selecting all the edges

Meshing

1. Horizontal and vertical edges are meshed into 50 divisions and 1 division respectively
2. Face mesh is performed

Boundary Type Specification

Edge	Name	Type
Left	Left wall	Wall
Right	Right wall	Wall
Top	Top wall	Wall
Bottom	Bottom wall	Wall

1. The continuum is selected as solid
2. The mesh file is exported with export 2D mesh option selected
3. The mesh file is saved as plate mesh
4. The meshed file is read to Fluent as case file
5. In the problem, the fluid is selected as water (L)
6. Aluminum has been selected as the material

Boundary Conditions

1. Velocity of 0.0003 m/s is applied to the left wall
2. The solution is initialized from the left wall
3. Then the solution is obtained by iterative process and is converged

Postprocessor

1. The graph is plotted
2. The solution is iterated for different time intervals
3. Comparative results are drawn

Inference

The flow over the cylinder is studied with contour representation of the pressure and velocity. The vertex formation at the downstream side cylinder is also studied.

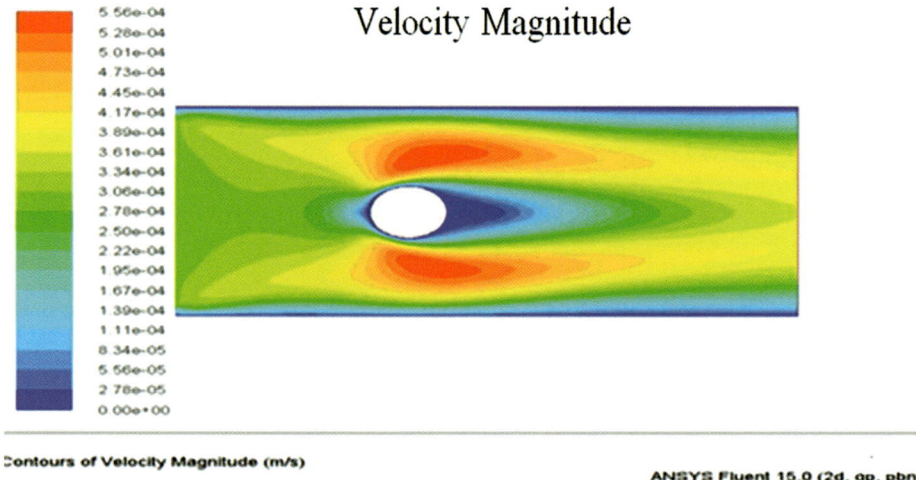

Contours of Velocity Magnitude (m/s)

ANSYS Fluent 15.0 (2d, op, pbns, lam)

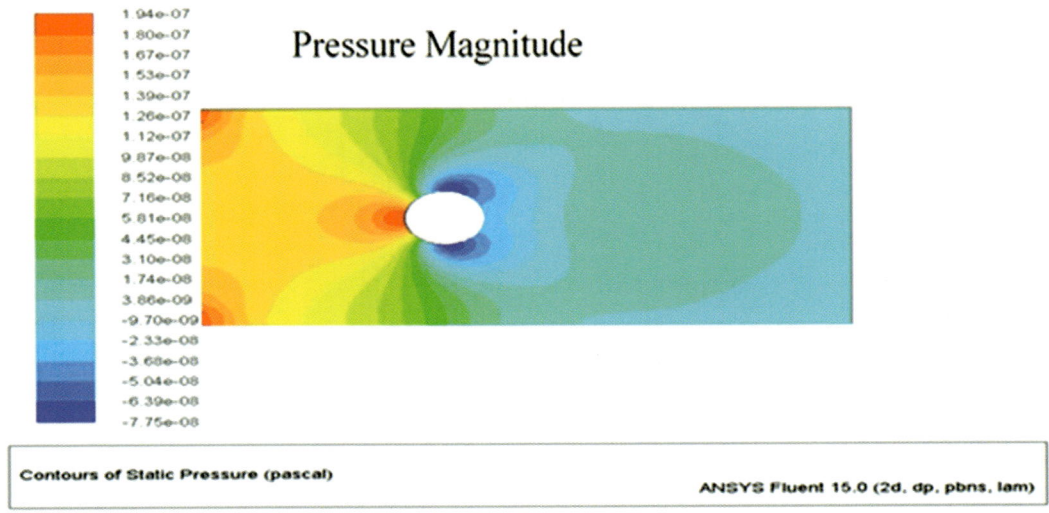

Contours of Static Pressure (pascal)   ANSYS Fluent 15.0 (2d, dp, pbns, lam)

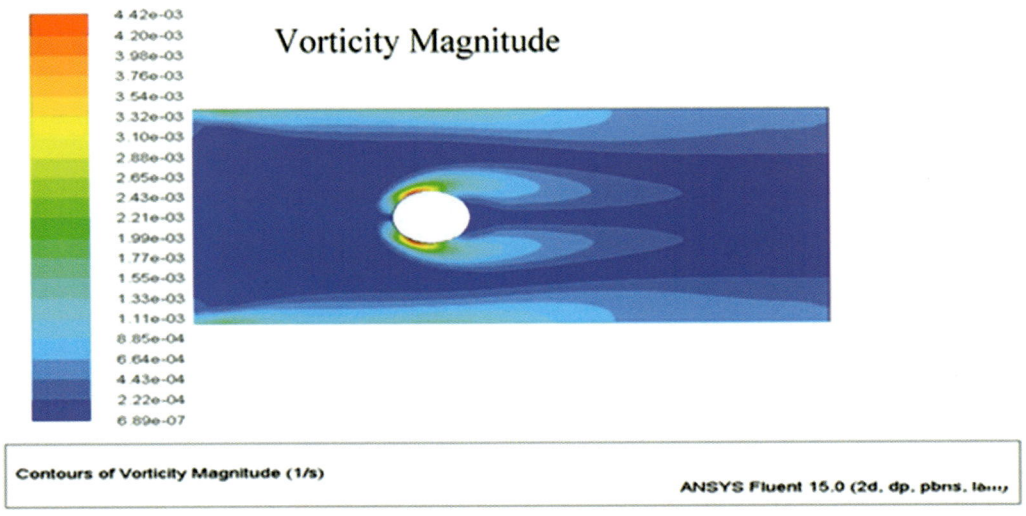

Contours of Vorticity Magnitude (1/s)   ANSYS Fluent 15.0 (2d, dp, pbns, lam)

RESULT

The problem is solved using Workbench and Fluent.

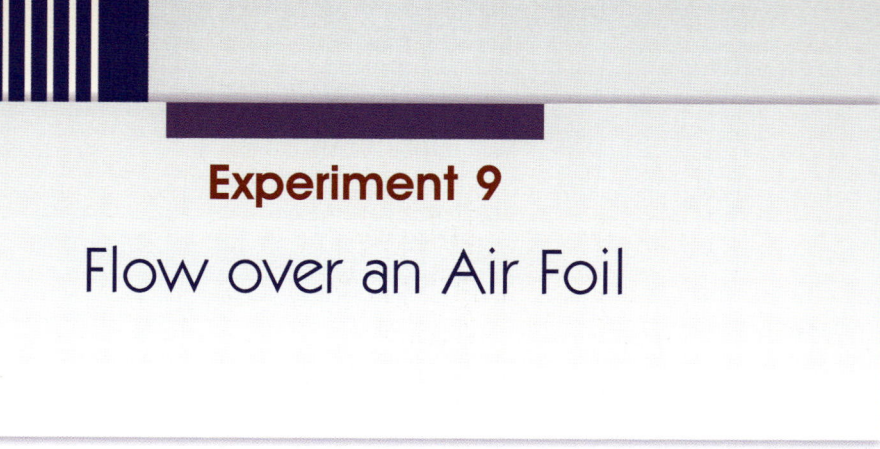

Experiment 9

Flow over an Air Foil

AIM

To solve flow over an air foil for a given problem using Workbench and Fluent.

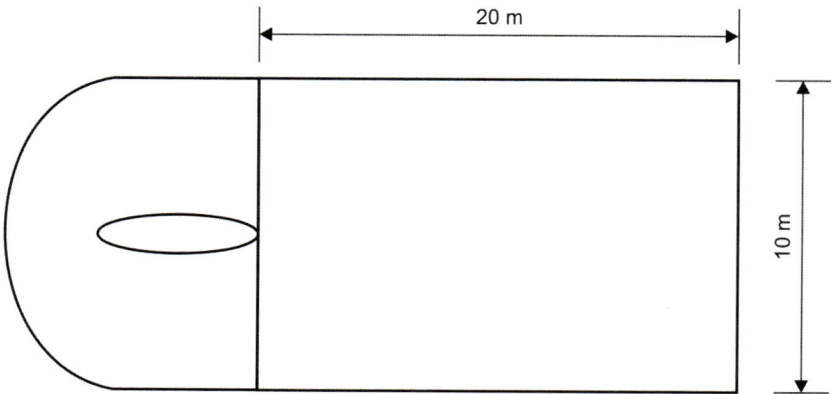

Problem Specification

- A plate of 10 × 20 m size and arc is drawn by using the sketch command arc by 3 point
- The air foil shape can be extracted from the file NACA series 0014

PREPROCESSOR

Model Creation

1. Edges are created using the vertices
2. Face is created by selecting all the edges
3. The Boolean option is used to subtract the airfoil surface from boundary

Meshing

1. Horizontal and vertical edges are meshed into 10 divisions and 1 division respectively
2. Face mesh is performed

Boundary Type Specification

Edge	Name	Type
Left	Velocity inlet	Wall
Right	Pressure outlet	Wall
Air foil	Wall	Wall

1. The continuum is selected as solid
2. The mesh file is exported with export 2D mesh option selected
3. The mesh file is saved as plate mesh

Boundary Conditions

1. The left wall is subjected to velocity of 1 m/s
2. The right wall is subjected to pressure as zero
3. Material properties are changed as given in the problem
4. The solution is converged by iterating process

Postprocessor

1. The graph is plotted between the length of plate on the x-axis and static temperature at the top wall on y-axis
2. The solution is iterated for different time intervals
3. Comparative results are drawn

Inference

The pressure and velocity distribution over the air foil is represented in the contours and the low pressure, high pressure and low velocity, high velocity regions are studied; also the lift convergence and drag convergence graph is plotted.

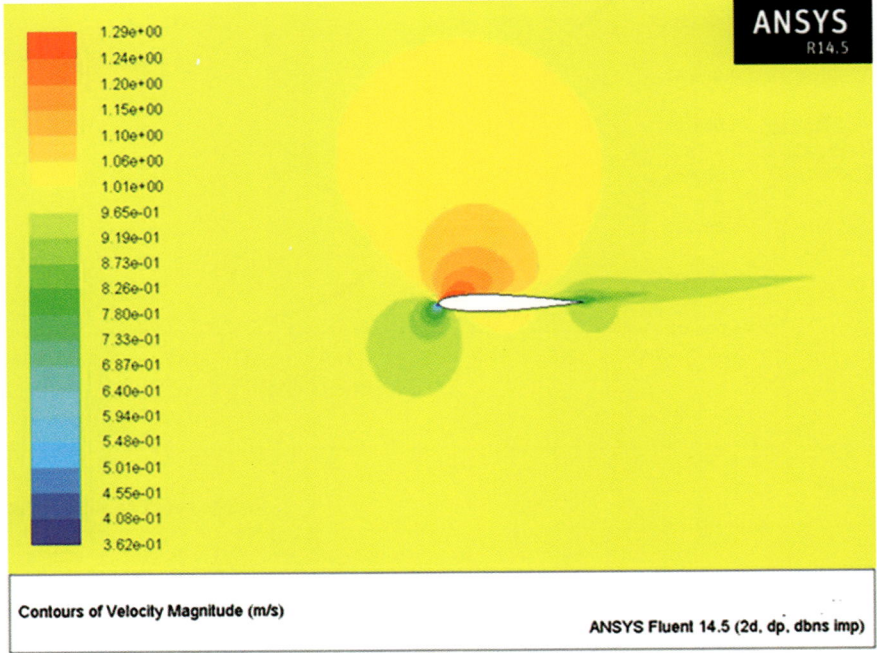

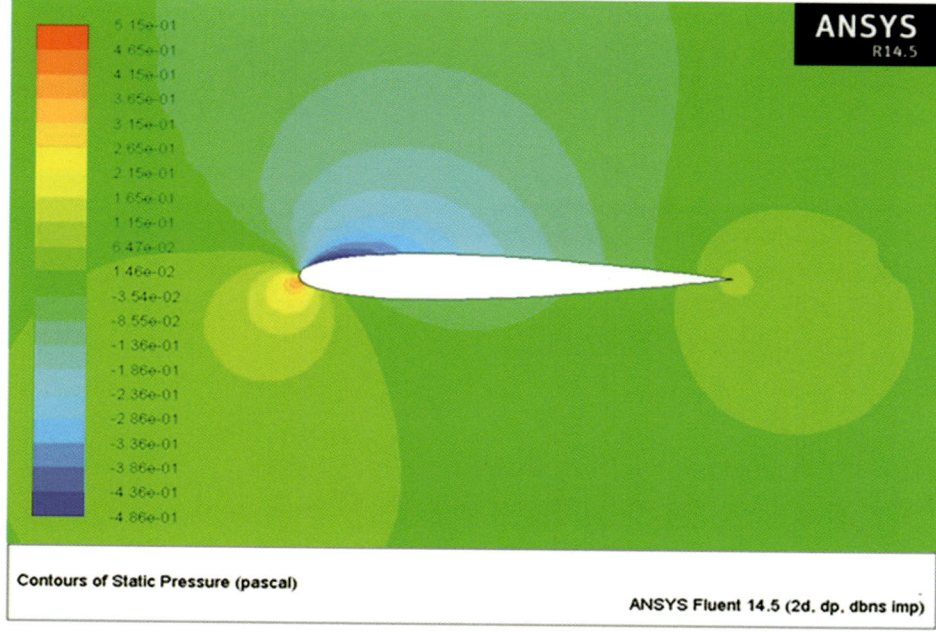

Contours of Static Pressure (pascal)

ANSYS Fluent 14.5 (2d, dp, dbns imp)

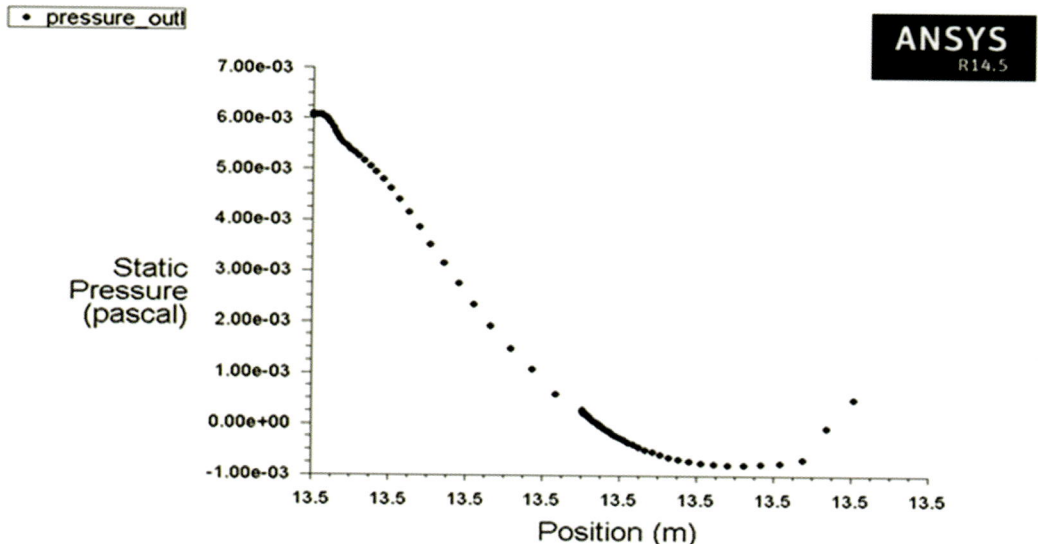

Static Pressure

ANSYS Fluent 14.5 (2d, dp, dbns imp)

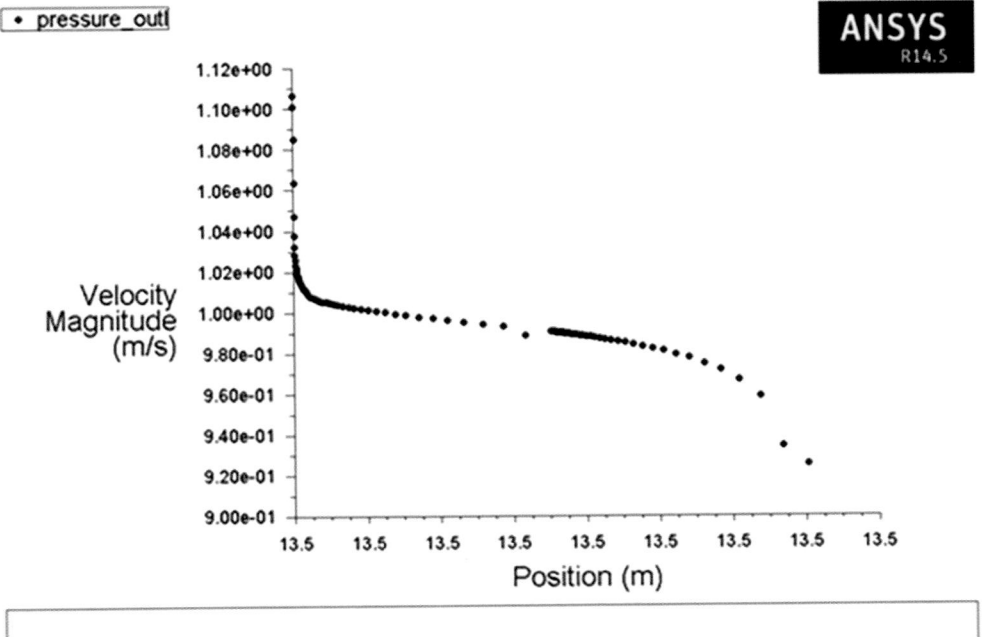

Velocity Magnitude

ANSYS Fluent 14.5 (2d, dp, dbns imp)

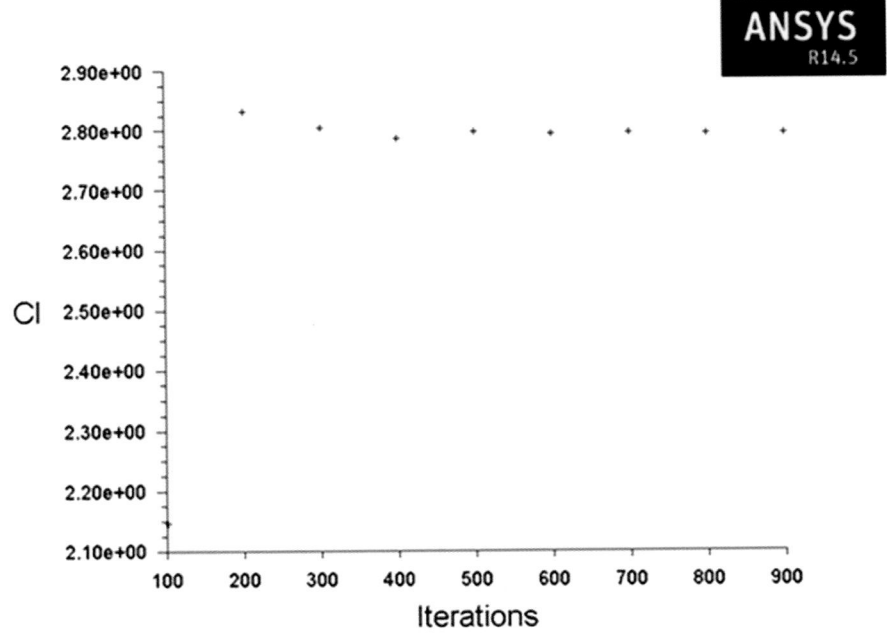

Lift Convergence

ANSYS Fluent 14.5 (2d, dp, dbns imp)

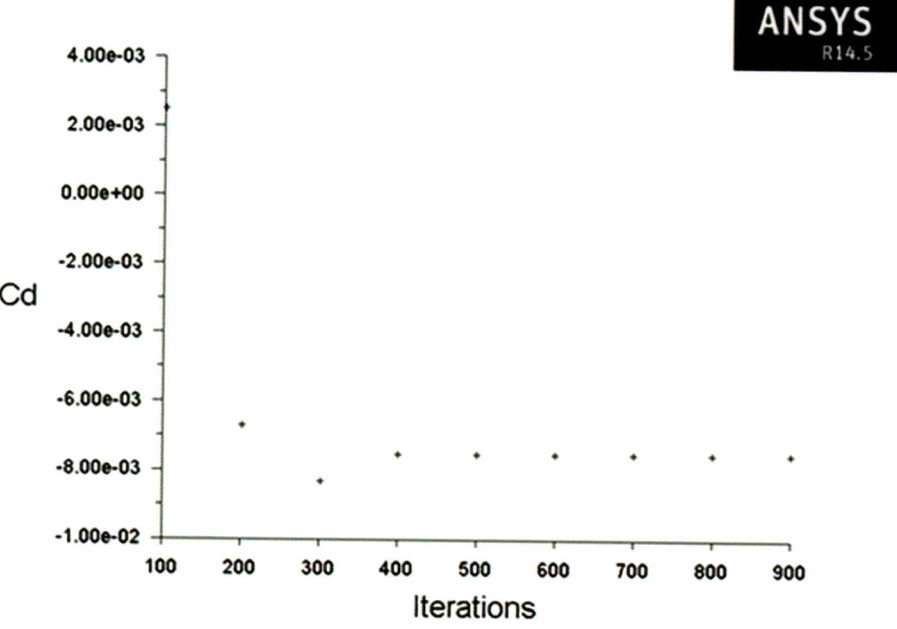

Drag Convergence

ANSYS Fluent 14.5 (2d, dp, dbns imp)

RESULT

The given problem is solved using Workbench and Fluent.

Experiment 10

Conjugate Heat Transfer

AIM

To solve conjugate heat transfer for a given problem using Workbench and Fluent to study the results.

Problem Specification

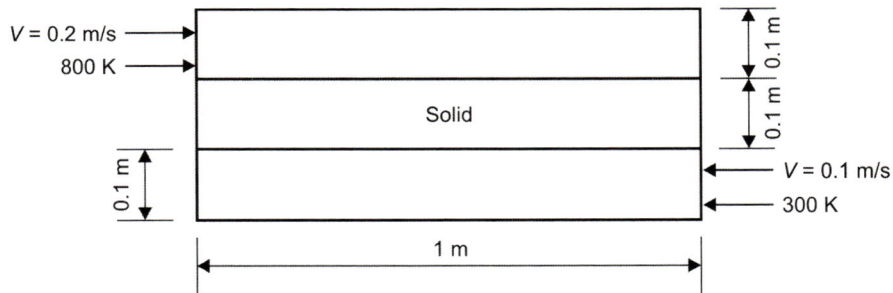

Properties of Fluid

Density	=	1000 kg/m³
Specific heat C_P	=	25 J/kg K
Thermal conductivity K	=	10 W/m K
Viscosity	=	0.15 kg/ms

Properties of Solid

Density	=	8000 kg/m³
Specific heat C_P	=	500 J/kg K
Thermal conductivity K	=	50 w/m K

PROCEDURE

Preprocessor

All the steps are carried out using Workbench

Model Creation

1. Vertices are created using (0,0), (1,0), (1,0.1) and (0,0.1), (0,0.2), (1,0.2), (1,0.3) and (0,0.3) coordinates
2. Edges are created using vertices
3. Faces are created using edges. Three faces are created

Meshing

1. Horizontal edges are meshed by 100 divisions
2. Vertical edges are meshed by 10 divisions each edge
3. Face mesh is performed

Specification of Boundary Type

Edge	Name	Type
Top left	Top Inlet	Velocity - Inlet
Top right	Top Outlet	Pressure - Outlet
Top	Top wall	Wall
Middle left	Left wall	Wall
Middle right	Right wall	Wall
Bottom right	Bottom inlet	Velocity - Inlet
Bottom left	Bottom outlet	Pressure - outlet
Bottom	Bottom wall	Wall

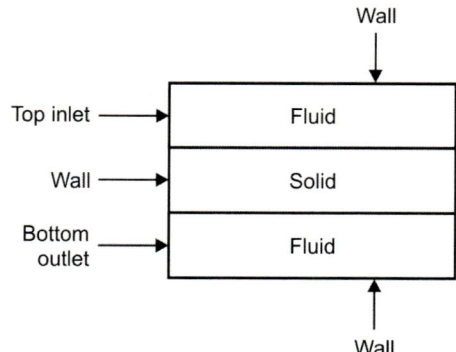

1. The continuum is selected as fluid and solid
2. The meshed file is exported with export 2D mesh option selected
3. The file has been saved as CHT.msh

Processor

1. The mesh file is read as a case file
2. Energy equation option is selected
3. The material properties for the fluid medium and the solid are changed as given in the problem

Boundary Conditions

1. Top inlet edge is applied at 0.2 m/s velocity and 800 K temperature
2. Bottom inlet edge is applied at 0.1 m/s velocity and 300 K temperature
3. Line has been drawn and (0.5, 0.3) value at that line, the temperature distribution has to be found out
4. Initialize with top-inlet
5. The solution is converged by iterative process

Postprocessor

The graph is plotted to the length on x-axis and static temperature at the line on y-axis.

Inference

1. Top and bottom curves are nonlinear, which is due to conduction and convection in fluid region
2. Middle curve is linear, because only conduction will take place in solid region

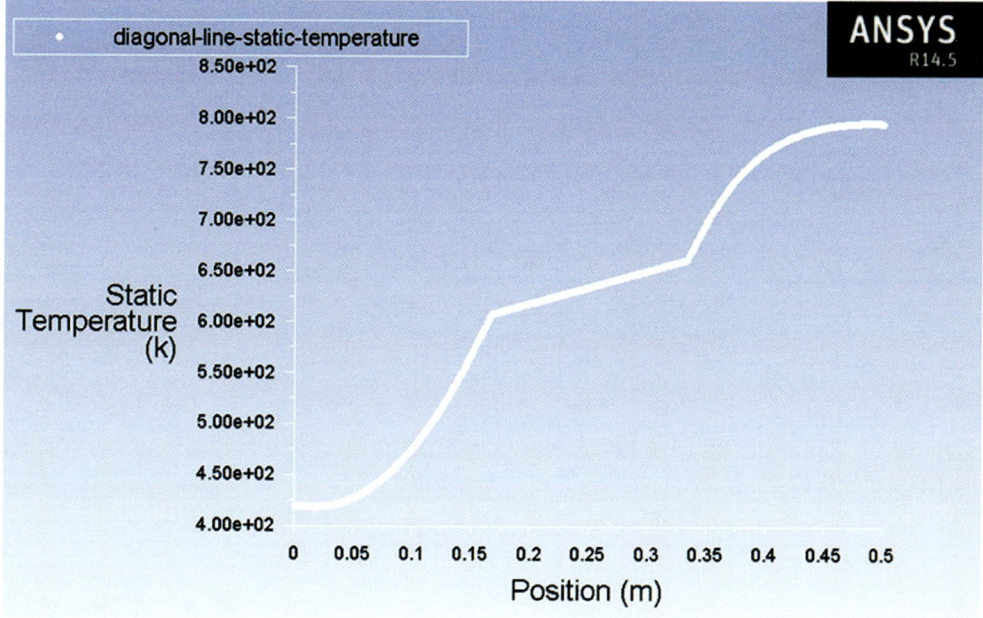

RESULT

The given problem is solved using Workbench and Fluent and the result is drawn.

READER'S NOTE

READER'S NOTE